SCIENTIFIC AMERICAN EDUCATIONAL PUBLISHING

EVERYDAY SCIENCE

10 FUN

EXPERIMENTS TO USE IN DAILY LIFE

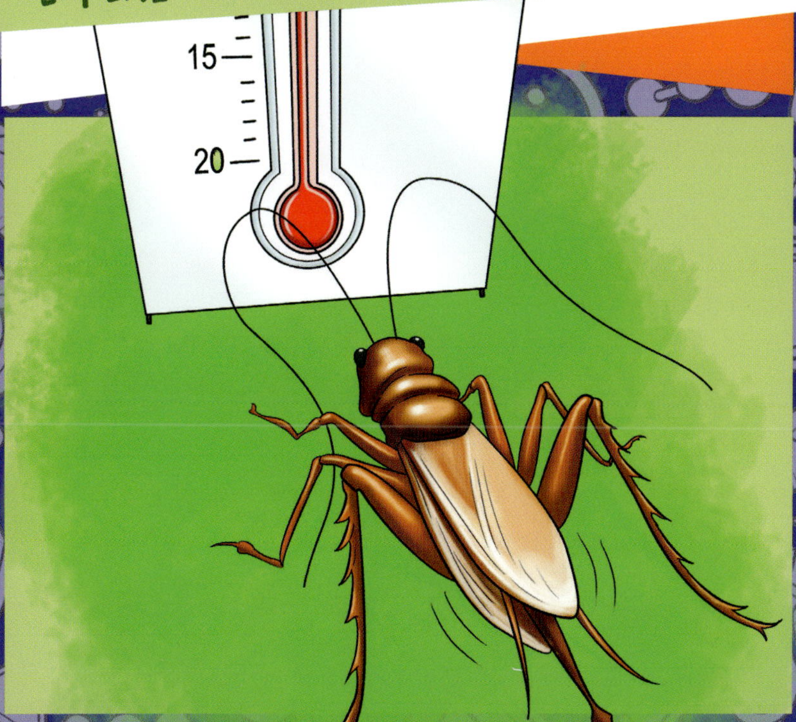

BRING SCIENCE HOME

Published in 2025 by Scientific American Educational Publishing in association with The Rosen Publishing Group.

Contains material from Scientific American, a division of Springer Nature America, Inc., reprinted by permission as well as original material from The Rosen Publishing Group.

First Edition

Editor: Kristen Rajczak Nelson
Designer: Rachel Rising

All illustrations by Continuum Content Solutions

Activity on page 5 by Science Buddies, Svenja Lohner (September 15, 2016); page 11 by Science Buddies (October 4, 2012); page 15 by Daisy Yuhas (July 5, 2012); page 18 by Ben Finio, Science Buddies (December 24, 2015); page 25 by Science Buddies (August 22, 2013); page 31 by Daisy Yuhas (September 20, 2012); page 37 by Science Buddies (December 8, 2011); page 43 by Ingrid Wickelgren (November 8, 2012); page 49 by Science Buddies (November 15, 2012); page 55 by Daisy Yuhas (November 22, 2012).

Photo Credits: pp. 3, 4, 5, 8, 11, 14, 15, 25, 28, 31, 33, 34, 49, 51, 55, 58 cve iv/Shutterstock.com; pp. 4, 5, 11, 15, 18, 25, 31, 37, 43, 49, 55 Anna Frajtova/Shutterstock.com.

Library of Congress Cataloging-in-Publication Data

Names: Scientific American, inc., editor.
Title: Everyday science : 10 fun experiments to use in daily life / Scientific American Editors.
Description: Buffalo, NY : Scientific American Educational Publishing, an imprint of Rosen Publishing, [2025] | Series: Bring science home | Includes index.
Identifiers: LCCN 2023056677 (print) | LCCN 2023056678 (ebook) | ISBN 9781725350557 (library binding) | ISBN 9781725350540 (paperback) | ISBN 9781725350564 (ebook)
Subjects: LCSH: Science--Experiments--Juvenile literature. | Scientific recreations--Juvenile literature.
Classification: LCC Q164 .E94 2025 (print) | LCC Q164 (ebook) | DDC 507.8--dc23/eng/20231226
LC record available at https://lccn.loc.gov/2023056677
LC ebook record available at https://lccn.loc.gov/2023056678

Manufactured in the United States of America

CPSIA Compliance Information: Batch #CSSA25. For further information contact Rosen Publishing, New York, New York at 1-800-237-9932.

CONTENTS

⚛ THESE ACTIVITIES INCLUDE SCIENCE FAIR PROJECT IDEAS.

INTRODUCTION

Science is all around you, and you use it every day! Whether you're trying to figure out how to remember something or entering a secure password into a website, science can help you to understand the world around you. These activities will help you to use science in your daily life.

Projects marked with ⚛ include a section called Science Fair Project Ideas. These ideas can help you develop your own original science fair project. Science fair judges tend to reward creative thought and imagination, and it helps if you are really interested in your project. You will also need to follow the scientific method. See page 61 for more information about that.

A Simple Sourness Detector

BUBBLE, BUBBLE, ACID, OR BASE? LEARN HOW YOU CAN EASILY FIND THE PH OF COMMON KITCHEN PRODUCTS—AND HOW THEIR ACIDITY RELATES TO THEIR TASTE.

Have you ever wondered why some foods or drinks taste sour whereas others do not? You might know that your tongue is picking up all kinds of flavors, including salty, sweet, bitter, and sour. But what makes something *taste* sour? The sour taste is actually influenced by the pH and acids present in foods. In this activity, you will find out how sour different foods are by testing foods and drinks for the presence of acids with baking soda. What do you think is the sourest food you can find in your kitchen?

PROJECT TIME

20 minutes

KEY CONCEPTS

Chemistry
Food Science
pH
Acidity

BACKGROUND

Taste is detected by taste buds that line the tongue and other parts of the mouth. The human tongue has an average of 10,000 individual taste buds! Inside each taste bud are as many as 100 receptor cells. Each receptor is best at sensing a single sensation and sends signals about it to the brain, which then identifies sweet, salty, bitter, sour, or umami. Different receptors are triggered by different chemical compounds (or molecules) present in the food. The sour taste receptors are triggered by acids, or more specifically hydrogen ions (H^+), and therefore are the tongue's acid detectors.

What does it mean for something to be acidic or basic? It all has to do with ions. In water (H_2O), a small number of the molecules split up to form hydrogen ions (H^+) and hydroxide ions (OH^-). Pure water and solutions that have an equal number of hydrogen ions and hydroxide ions are neutral—that is, they are neither acidic nor basic. An acid is a substance that donates hydrogen ions. Because of this, in an acidic solution there are more hydrogen ions than hydroxide ions. This means the more acidic a food is, the more hydrogen ions are available to trigger the sour taste receptors. A base is a substance that accepts hydrogen ions. When a base is dissolved in water, the balance between hydrogen ions and hydroxide ions shifts the opposite way so the hydroxide ions dominate. The acidity of a solution is measured by the pH scale. A pH of 7 is neutral; everything with a pH below 7 is acidic whereas substances with a pH above 7 are basic, or alkaline.

But there is a simpler way besides measuring pH to find out if a food is acidic or not—by using baking soda. Also known as sodium bicarbonate, it is a base that reacts with acids to generate carbon dioxide gas, a reaction well known from science fair volcanoes or bottle rockets. The escaping gas can be easily spotted as it forms bubbles once the acid comes in contact with the baking soda. This makes the "sour test" simple: The more gas bubbles you see, the more acidic your food is—and the sourer it should taste. Test it out for yourself!

MATERIALS

- Six cups or glasses
- Plate
- Vinegar
- Soda/soft drink
- Lemon juice
- Milk
- Dish soap
- Water

- Spoon
- Baking soda (It must be fresh; if it is too old, it will not work as well.)
- Medicine dropper or pipette
- Fruits or fruit juices and other foods or drinks
- Baking powder (optional)

PREPARATION

- Take one cup and fill it with tap water. Then add a squirt of dish soap to make a solution for testing. **Note:** This is the one solution that you should not taste; the rest will be okay to sample.
- Pour a little bit of the other test solutions (water, lemon juice, vinegar, milk, soda) each into its own cup.
- Use the spoon to cover the plate evenly with baking soda.

PROCEDURE

- Start with the vinegar first. Take the medicine dropper or pipette and suck up a little of the vinegar from the cup.
- Bring the dropper to the plate with baking soda and slowly squirt one single drop onto the baking soda on the plate. *What happens once the vinegar drop hits the baking soda? Can you see some bubbles?*
- Rinse your dropper with some water and then use it again to take a sample of the soft drink. Again, squirt one drop carefully onto the baking soda. Choose a different spot on the plate. *Is there a reaction happening? Do you see carbon dioxide generation? How does the reaction compare with the vinegar?*
- Rinse the dropper again and move on to the next sample: Take a little bit of water and put one drop of your sample onto the plate. *What happens this time? Is water acidic?*

- With a clean dropper, suck up some lemon juice and squirt a drop onto the baking soda. Choose a fresh spot that does not have any solution on it yet. *How many bubbles do you see? Did it react more than the water or less?*
- Repeat the same procedure with the milk. *Does the baking soda react with the milk to form bubbles? Do you think milk tastes sour?*
- Test your last sample, which is the dish soap solution. *Can you tell from the reaction if it contains a lot of acids? What pH do you think the soap solution has?*
- *From all your observations, can you make predictions about which of the liquids contains the most acids or would taste the most sour?* You can repeat the baking soda test on all of your samples once more. *Which one is the most acidic? Which one is the least acidic?*

EXTRA

Do some research to find out the pH of all your test solutions. *Do your results agree with your findings?*

SCIENCE FAIR IDEA

Can you find more liquids to test? You can also use different fruits or vegetables and squeeze them to get some liquid to test if they are acidic or sour. Do a taste test afterward to see if your observations match your taste.

SCIENCE FAIR IDEA

Repeat the same procedure but this time use baking powder instead of baking soda. Do some research to find out what the difference between the two is and compare the reactions with the same test solutions. *Do you see more bubbles with the baking soda or with the baking powder for each of the same samples?*

EXTRA

Instead of putting a drop of your test solution onto the baking soda, try to do the same test the other way around. Take a little bit of the baking soda and put it into the glass with your test solution. *What happens when the baking soda comes in contact with the liquid? Is the reaction the same as when you put the solution onto the baking soda or different?*

EXTRA

If you have access to litmus paper (which is used to measure the pH of a solution), measure the pH of each of your test samples. *What is the pH of each of them? Did all of the acidic samples form bubbles during your baking soda test?*

OBSERVATIONS AND RESULTS

Did you get some nice bubbles for some of your test solutions? You should have seen a vigorous bubble formation with lemon juice as well as with vinegar. They both contain lots of acids. The juice of a lemon contains about 5 to 6 percent citric acid, which also gives the lemon its sour taste. Vinegar consists of 5 to 20 percent acetic acid, which is the acid that reacts with baking soda to form carbon dioxide. With the soft drink, you also should have seen some bubble formation, although not as much as with lemon juice and vinegar. The reason for an acidic reaction is the phosphoric acid that is added to soft drinks for preservation and taste enhancement. This is one reason why it is so bad for your teeth (besides the sugar); this acid can damage the tooth enamel.

Milk is a pretty neutral solution—similar to pure water—and should not have reacted with the baking soda. Soap, on the other hand, is a basic solution, which means its pH is greater than 7. It contains no acids that can react with the baking soda and should not have resulted in carbon dioxide formation. When you used baking powder instead of baking soda, however, you might have noticed that the reactions for each of your test samples were more pronounced. You probably even saw some tiny bubbles with the solutions that did not react with the baking soda. This is due to the fact that baking powder is a mixture of baking soda and weak acids. When the baking powder gets wet, these acids can react with the baking soda and form carbon dioxide.

When doing a taste test, you might have noticed that some foods or drinks did not taste sour at all—even if they contain lots of acids. The trick is that the sour taste is often masked with the addition of lots of sugar. One good example is the soda drink that you tested. Look at its ingredient list—how much sugar does it have? Do you think you would still like it without the sugar?

CLEANUP

Pour the test solutions down the drain of your kitchen sink. You can dispose of the baking soda in your trash or rinse it off with water into the sink.

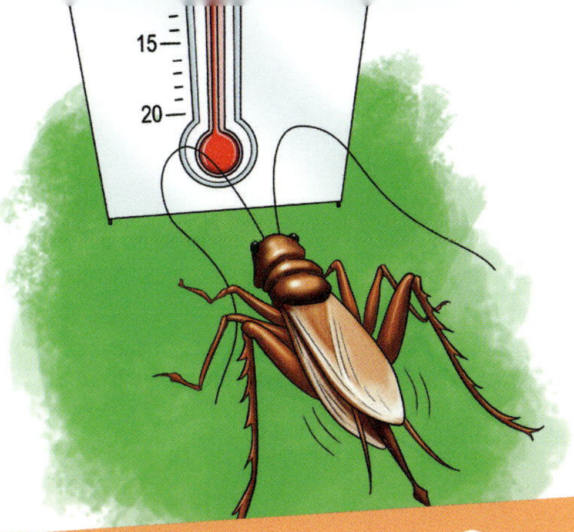

Sonorous Science

Have a Cricket Tell You the Temperature!

DID YOU KNOW THAT CRICKETS ARE ALSO THERMOMETERS? TRY THIS EXPERIMENT TO SEE HOW A BUG CAN TELL YOU HOW WARM OR COLD IT IS.

Do you love to hear the pulsing chirp of crickets as you fall asleep? It is an unmistakable sound. Some people find the sound pleasing and peaceful whereas others find it loud and annoying, especially if a cricket happens to find refuge from the cold inside a home. However you feel about crickets, their chirps hold a surprising message—they can report the temperature! In this activity, you'll investigate how the chirps of these tiny creatures can be used as a kind of thermometer!

PROJECT TIME

20 minutes

KEY CONCEPTS

Temperature
Mathematics
Energy
Chemical reactions
Insects
Sound

BACKGROUND

How is a cricket's chirp related to temperature? Crickets, like all living things, have many chemical reactions going on inside their bodies, such as reactions that allow muscles to contract to produce chirping. Crickets, like all insects, are cold-blooded and take on the temperature of their surroundings. This affects how quickly these chemical muscle reactions can occur. Specifically, a formula called the Arrhenius equation describes the activation, or threshold, energy required to make these reactions occur. As the temperature rises, it becomes easier to reach a certain activation energy, thereby allowing chemical reactions, such as the ones that allow a cricket to chirp, to occur more rapidly. Conversely, as the temperature falls, the reaction rates slow, causing the chirping to diminish along with it.

How do crickets make their distinctive chirp? They use a process called stridulation, where special body parts are rubbed together to make a noise. Generally, only male crickets do this; there's a special structure on the tops of their wings called a scraper. When they want to make their sound, they raise their wings to a 45-degree angle and draw the scraper of one wing across wrinkles on the underside of the other wing, called a file. It's somewhat like running your finger along the teeth of a comb.

MATERIALS

- Outdoor temperature between 55 and 100°F (12 and 38°C); during an evening is ideal
- Access to an outdoor area with crickets or purchase crickets from a pet store
- Outdoor thermometer
- Stopwatch
- Piece of scratch paper and pencil or pen

PREPARATION

- Set up a thermometer to measure the outdoor temperature in the area where you will observe the crickets.
- While it's between 55 and 100°F (12 and 38°C) outside (and keeping in mind that evenings are usually best for hearing crickets chirping), go to the area where crickets are. Make sure you hear some chirping.

- Alternatively, if you purchase crickets, set them outside in a cage in the shade that allows the outside air to easily reach them. Wait until you hear chirping.

PROCEDURE

- Pick out the chirping sound of a single cricket.
- Count how many chirps the cricket makes in 14 seconds. *How many?* Write this number down.
- Do this two more times, counting how many chirps the cricket makes in two more 14-second intervals. Write these numbers down. *How close were the numbers to one another?*
- Average the number of chirps in the 14-second intervals.
- Add 40 to the average number of chirps in 14 seconds. This equation (one of the easiest to use cricket-thermometer equations) is published in the *Farmers' Almanac*. It should give you the approximate temperature in degrees Fahrenheit. *According to the cricket, what is the temperature?*
- Check the temperature on the outdoor thermometer. *How close is the temperature based on the cricket chirps to the thermometer reading? If they are different, by how much are they different? Why do you think they might be different?*

EXTRA

To see how accurate it is to use a cricket to tell the temperature in general, repeat this activity with different outdoor temperatures (but stay within the 55 and 100°F [12 and 38°C] range). *How accurate is the cricket at telling the temperature when it is getting warmer or colder outside? How accurate does it seem overall?*

EXTRA

Based on the *Farmers' Almanac*, you can use a cricket to tell the temperature in degrees Celsius by counting the number of chirps in 25 seconds, dividing this number by 3 and then adding 4. *According to the cricket, what is the temperature in degrees Celsius? How accurate is this?*

SCIENCE FAIR IDEA

Compare the chirps of different species of crickets or different insects altogether, such as katydids. *Which critter makes the best "insect thermometer"?*

OBSERVATIONS AND RESULTS

Was the temperature based on the cricket's chirps close to the actual outdoor thermometer temperature, and maybe off by only a few degrees Fahrenheit?

As far back as the late 1800s, there have been articles published noting that a cricket's chirp rate (or number of chirps per second that it makes) changes consistently based on the outdoor temperature. There have been many equations published describing the relationship between the number of chirps per second and the temperature. These equations all vary slightly, depending on the species of cricket. Using this activity, you may have found that the cricket was within about 5 degrees Fahrenheit of the temperature measured using the outdoor thermometer, and probably even closer than that. If you repeated this activity multiple times and found that, based on the cricket's chirps, it was colder than the thermometer read, this could be because the cricket was farther away from a warm building than the thermometer was, and/or because the cricket was closer to the cold soil. The snowy tree cricket is frequently cited as the most accurate at predicting temperature.

CLEANUP

Put away all supplies.

Sound Science

Where Did That Noise Come From?

LEARN ABOUT YOUR SENSE OF HEARING AND COMPARE HOW WELL EACH OF YOUR EARS CAN HEAR!

In the popular pool game "Marco Polo," one sightless swimmer calls out "Marco" and seeks out other swimmers when they respond "Polo." If you've tried it, you'll know it's not an easy game. Every splash, whisper, or giggle is a clue, but these hints can be surprisingly misleading. In this activity, you'll learn a little bit more about your sense of hearing and how it works.

PROJECT TIME

15 minutes

KEY CONCEPTS

The five senses
Neuroscience
Sound waves
Audition

BACKGROUND

Most of us chiefly rely on eyesight, which is part of why we become so disoriented when forced to resort to a different sense. Nevertheless, our ears share a lot of important information about the world around us. From the shout, "Fire!" to a relaxing melody, your ears convey useful signals, and you wouldn't want them to deceive you.

Sound travels through the air in waves. Your ears are specially equipped to receive and understand these waves. Each ear collects and channels sound waves, transforming them into vibrations. Within your inner ear, tiny hair cells respond to these vibrations and send signals that your brain can decode and interpret as a variety of sounds. But why exactly do we have two ears instead of just one? Try this activity and find out.

MATERIALS

- Adhesive tape (or chalk if outside on pavement or sidewalk)
- Meter stick or measuring tape
- Blindfold
- Earplugs (optional)
- One or two noisemakers: a clock that audibly ticks, music box, small musical instrument, cell phone, pennies in a closed jar, a book (optional)
- Pencil and paper

PREPARATION

- Use the tape (or a piece of chalk) to make an X in the center of the room or outdoor area.
- Make lines around the room or area using tape at varying distances from the X. Make some close and some farther away.

PROCEDURE

- Blindfold your partner and ask them to stand on the central X. Record or draw the direction your partner is facing and ask your partner to remain in that spot throughout the activity.

- Explain to your partner that you will stand in different parts of the room and say their name. Your partner will then guess where you are (left, right, in front, behind) and how far away you are standing.
- When you're both ready to begin, ask your partner to cover one ear using a hand or by inserting an earplug.
- Stand on one of the lines you have made in the room. Say your partner's name in a normal speaking voice. Ask them to guess where and how far away you are standing. *Did your partner guess where you were standing?*
- On a piece of paper, record which ear your partner used, where you actually stood (distance and direction from the central X), and what your partner guessed.
- Move to a different location and repeat. Try this from different lines all over the room or area and record your partner's guesses. *Is your partner better at guessing when you're close by or far away? How well does your partner guess distance? What about direction?*
- Once you've tried several spots in the room, ask your partner to use both ears and try the above steps again. *Is your partner better at guessing distance with two ears than one? What about direction?* You can also try the above steps after asking your partner to cover the other ear. *Is one ear better for guessing locations and distances than the other? Does it make any difference which ear is covered?*
- Change places with your partner and compare your findings!

EXTRA

Try this activity and vary how loudly you say your partner's name. *Does volume make it easier or more difficult for your partner to judge distance? What about direction? Is a whisper harder to locate than a shout?*

EXTRA

Instead of speaking, try using an object that makes noise such as a ticking clock, music box, or cell phone—or use a homemade noisemaker by shaking pennies in a closed jar or opening and sharply closing a hardbound book. *How does a noisemaker compare with a voice? Are some sounds harder to locate than others?* Try this activity while your partner stays perfectly still, then allow your partner to move his or her head. *Does head turning or tilting make a difference?*

OBSERVATIONS AND RESULTS

Were two ears better than one in identifying distance and direction? Was direction more difficult to guess with one ear when you and your partner were closer to one another?

Each ear receives information that is sent to your brain. Because your ears are not side by side, they receive different information. If someone standing to your left claps their hands, your left ear will receive this sound wave more quickly than your right one. In addition, the clap will sound louder in your left ear than in your right. Your brain uses these differences to better understand where a sound is coming from. This can also explain why—as you may have noticed—it's hard to tell the difference between a sound directly in front of or behind you, even if you are using both ears. When the sound source is exactly equidistant to both ears, they receive very similar information and your brain has fewer clues as to where the source may be.

CLEANUP

If you are indoors, remove the tape from the floor. Return all materials to the places where you found them.

Password Hacker

LEARN HOW PASSWORDS CAN BE MADE HARDER TO GUESS—WHILE PLAYING A FAST-PACED GUESSING GAME. SEE IF YOU CAN CRACK THE CODE!

Do you have your own email or other online account? If so, you probably use a password to log in. How did you pick your password? Is it something that might be easy for someone else to guess, like the name of your pet? This fun activity will teach you about password security and how to pick a good password.

PROJECT TIME

10 minutes

KEY CONCEPTS

Technology
Cybersecurity
Internet safety
Privacy

BACKGROUND

Imagine a suitcase lock with three number wheels on it. Each wheel contains digits 0 through 9, so you can pick any three-digit number as the combination. Because each wheel has 10 digits, there are 10 X 10 X 10 = 1,000, or 10^3, possible combinations. If you add a fourth wheel, now there are 10 X 10 X 10 X 10 = 10,000, or 10^4, possible combinations. What if you used the letters A to Z instead of numbers 0 through 9 for the three-wheel lock? Then there would be 26 characters for each wheel, so there would be 26 X 26 X 26 = 17,576, or 26^3, possible combinations. Adding more digits, or more characters per digit, greatly increases the number of possible combinations. This makes it very difficult for a human to guess the combination.

However, computers can try to guess passwords much faster than a human can guess a physical lock combination. A fast computer can try to guess *millions* of passwords per second. That is why passwords should usually be long—at least eight digits—and made up of different types of characters including numbers, symbols, and upper and lowercase letters. A short password made only of lowercase letters might be very easy for a computer to guess. This activity will demonstrate how shorter passwords made up of fewer types of characters are easier to guess than longer passwords with more characters. Since humans will be doing the guessing, the passwords will be very short (just one or two digits) but remember that real-life passwords should be much longer.

Also note that there are other general password safety rules that you should follow. Just because a password is long and mixes letters and numbers does not mean it won't be easy to guess. For example, your name followed by your birthday could be easy for someone who knows you to guess. There are also many commonly used passwords like "password," "qwerty," or "123456789" that you should avoid.

MATERIALS

- A friend or family member
- Pen or pencil and paper (optional)

PREPARATION

- In this activity, you will compete in a series of password "duels" against an opponent. The catch: each person will have to follow different rules for thinking of a password. The rules determine how many total possibilities there are for that person's password. *Will passwords with more total possibilities be harder to guess?*

PROCEDURE

- To start, tell your opponent that you will think of a number between 0 and 9. Tell your opponent to think of a letter from A to Z. These are your "passwords." *Which type of password do you think will be harder to guess?* (Remember that real-life passwords are much longer. The passwords in this game are very short so the activity doesn't take too long.)
- Now, take turns trying to guess the other person's password. You can decide who gets to go first. Then, keep alternating until one person's password is guessed. *Whose password was guessed first? How many guesses did it take?*
- Repeat the same duel four more times, for a total of five rounds. Each time, you should think of a number from 0 to 9, and your opponent should think of a letter from A to Z. The rounds are totally independent, meaning it is okay to re-use the same password if you want. *Does one type of password seem easier to guess than the other?*
- Now, switch roles. You think of a letter from A to Z, and your opponent thinks of a number from 0 to 9. Repeat for a total of five rounds. *Which type of password gets guessed more frequently?*
- Now, change the rules for the passwords. You will think of a letter from A to Z, and your opponent will think of a number from 0 to 100. *How many possibilities are there for each type of password?*
- Repeat for a total of five rounds of the game, then switch roles and do five more rounds. *Which type of password is harder to guess? Does having more total possibilities make a password harder or easier to guess?*

EXTRA

Keep a tally mark of how often each type of password is guessed and make a graph of your results. *Which type of password is guessed the most often? The least often?* To get enough data for a good graph, you might need to do more duels with your opponent or get other people to join and collect all the data.

EXTRA

Try the activity with other rules for passwords. *For example, what if someone is allowed to pick a number 0 to 9 or a letter A to Z? What about a two-digit password made of numbers or letters (for example, "A7")?* Pit different combinations of password rules against each other in duels and keep track of all your results.

OBSERVATIONS AND RESULTS

You should find that passwords with fewer total possibilities are easier to guess than those with more possibilities. For example, in the first matchup where one person thinks of a number 0 to 9 and the other person thinks of a letter A to Z, the person with a number 0 to 9 will usually (but not always) "lose" the duel. Depending on how quickly you guess back and forth, each duel should take less than a minute.

Because of the random nature of guessing, it is important that you do enough duels to see this trend, which is why we suggest doing at least 10 of each type of duel. If you only do a couple duels, there is a higher chance that one person will "get lucky" and guess the other person's password, even if that person has the more difficult password type (as analogy, think about flipping a coin: if you only flip a coin twice, there is a relatively high chance that you will get two heads or two tails. However, if you flip a coin 1,000 times, your results should be very close to 50:50).

You should find that very short (one or two digit) passwords work best with this activity. Longer passwords (three digits or more) have so many possibilities that they generally take a very long time for a human to guess.

CLEANUP

If you used paper and a pencil for this experiment, put them away.

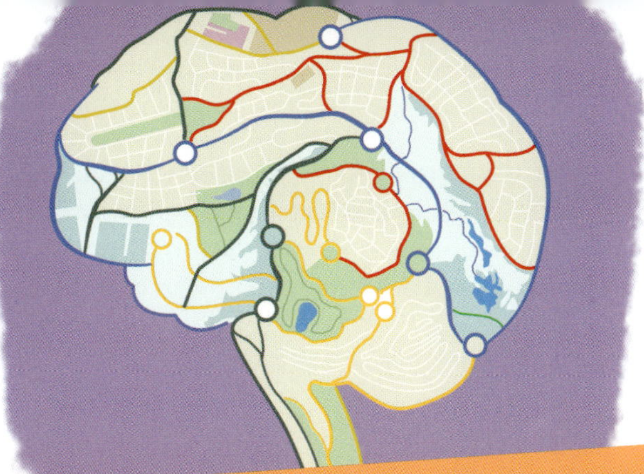

Seeing Science

Solving the Mystery of the Shrinking Moon

IS YOUR PERCEPTION OF THE MOON'S SIZE CORRECT? TRY THIS EXPERIMENT AND FIND OUT!

Have you ever noticed how the moon appears much bigger at the horizon as it is rising over the nearby buildings or treetops, than it does later in the evening when it is directly overhead? Of course, the moon's size does not change, but our perception of its size changes based on its position in the sky. In this activity, you'll investigate Emmert's law, which helps explain the full-moon illusion, and estimate the size of the perceived increase in size of the moon when it is near the horizon. Then you could check out the real moon and see how this activity holds up to the full-moon illusion.

PROJECT TIME

20 minutes

KEY CONCEPTS

Perception
Optical illusions
Human biology
Space

BACKGROUND

A full moon rising over the horizon often seems to be unusually large, but looks smaller as it moves up in the sky. The actual size of the moon stays the same. So what is the basis for this illusion? One well-supported theory is that the brain "thinks" the sky overhead is closer than the sky at the horizon and it adjusts the size of the moon's image accordingly. When the moon is near the horizon, your brain miscalculates the moon's true distance and size, making it seem larger in relation to its surroundings.

One way to explore this illusion is with afterimages. These occur when certain light-sensing cones in your eyes become fatigued after staring at a bright—or brightly colored—object. You can manipulate afterimages to mimic your perception of the moon at different places in the sky. Although the actual size of a specific afterimage on your retina doesn't change, its perceived size can, depending on your perception of how far away the surface on which you view it seems to be. This phenomenon is known as Emmert's law.

MATERIALS

- A sheet of blue construction paper (A blue pen or pencil can be used instead, but the colored paper is best.)
- Scissors
- Glue or tape
- A sheet of yellow construction paper (A white sheet of paper can be used instead but is less preferable.)
- A timer or a clock that shows seconds
- A helper
- An area with a clear view of the horizon and the zenith (the region of the sky directly overhead) (This part of the activity should be done mid-morning or mid-afternoon—to avoid looking at the sun—and on a day that is fairly cloudless with a lot of blue sky.)

PREPARATION

- Cut a square out of the blue construction paper sheet, about 1 to 2 inches (2.5 to 5 cm) on each side.
- Lay the yellow sheet of construction paper down in the landscape position (longways horizontal). Fold the paper in half (taking the right side of the paper and folding it over the left side). Then unfold the paper.
- Using glue or tape, attach the blue square to the yellow sheet of construction paper, in the middle of either the left or right half of the yellow sheet. Be careful not to cover the top of the square or yellow sheet with the tape or glue.
- If you do not have construction paper, you may draw a solid blue square on either the right or left side of a plain white sheet of paper, as described.

PROCEDURE

- You can complete the first part of this activity inside: hold the yellow paper with the blue square in front of you. Stare at the blue square for 30 seconds. (Use a timer with a buzzer or have a helper watch a clock for you.) Without changing the distance between your head and the yellow paper, switch from looking at the blue square to looking at the empty half of the yellow paper. *Do you see the afterimage of the square?* If you cannot clearly see an afterimage, try repeating this step until you can.
- Once the afterimage fades, keep your head the same distance from the paper as it was before and again stare at the blue square for 30 seconds. Then look for the afterimage on the empty half of the yellow paper. But this time, try moving your head away from or closer to the yellow paper. *How does the size of the afterimage change as you change the distance between your head and the yellow sheet?* If you could not clearly see a change, try repeating this step until you can.

- *Overall, how did the size of the afterimage change as you changed the distance between your head and the yellow paper?*
- Next, go outside to an area with a clear view of the horizon and zenith. Do not do this at noon, when the sun is directly overhead or at twilight when the sun is low on the horizon, because this interferes with your observation of the afterimage at the zenith (the sky directly overhead) or the horizon. Mid-morning or mid-afternoon is probably the best time to do this, and the day should be fairly cloudless with a lot of blue sky to look at. Remember: when choosing the section of the sky to use, be sure that it does not coincide with where the sun is—you should never look directly into the sun.
- Stare at the blue square on the yellow sheet for 30 seconds and then look at the horizon. *Is there an apparent change in the size of the afterimage? In other words, does the afterimage look smaller, larger, or the same size as the blue square?* You can repeat this step a few times if you are unsure of your observations.
- Next, stare at the blue square for 30 seconds and then look at the zenith. *Is there an apparent change in the size of the afterimage? Does the afterimage look smaller, larger, or the same size as the blue square?* Again, you can repeat this step if you are unsure of your observations.
- *Overall, did the afterimage appear smaller, larger, or the same size at the horizon compared with its appearance at the zenith?*

SCIENCE FAIR IDEA

Try the first part of this activity again (before you go outside), but this time try to quantify how the size of the afterimage changes depending on the distance between your head and the yellow sheet. (Be sure to always stare at the blue square from the same distance.) You could put the yellow sheet on a wall next to a ruler to estimate the size of the afterimage. Alternatively, you could draw several different size squares (some bigger and some smaller than the blue square) nested together on the yellow sheet and try to see in which square the afterimage fits best. *How does the distance between your head and the afterimage quantitatively correlate to the size of the afterimage?*

EXTRA

Try to estimate the change in the size of the afterimage at the zenith compared with the horizon. To do this you could cut out several squares (a white sheet of paper would work), some smaller and some larger than the blue square, and hold them out at arm's length when looking at the afterimages. Try to find the squares that are most similar in size to the afterimages. *Based on your findings, what is roughly the magnitude of the change of the full moon's apparent size between the horizon and the zenith?*

EXTRA

Try to estimate the change in the size of the afterimage at 45 degrees above the horizon; that is, halfway between the horizon and the zenith. (Be sure to avoid the part of the sky with the sun.) *What is the relative size of the afterimage at 45 degrees above the horizon, and how does it compare with the afterimage's sizes at the zenith and the horizon?*

OBSERVATIONS AND RESULTS

When you stared at the blue square, and then looked at the yellow sheet and moved your head away from or closer to the sheet, did the size of the afterimage increase with distance? Did the afterimage look bigger on the horizon compared with its size at the zenith?

Whereas the actual size of the afterimage on your retina doesn't change, the perceived size of the afterimage actually grows as you increase the distance between you and the surface on which you view the afterimage. (Or, in other words, the perceived afterimage size decreases as you get closer to the surface you're viewing it on.) This phenomenon is known as Emmert's law. One theory for why we perceive the full moon to be larger at the horizon compared with the zenith is that we perceive the sky overhead, at the zenith, as being closer than the sky at the horizon. In this activity, this should have been apparent using afterimages; the afterimage at the horizon should have appeared larger than the afterimage at the zenith (although maybe only by a little, such as approximately 1.3 to 1.5 times larger at the horizon, depending on the exact conditions).

CLEANUP

Recycle any paper you used. Return the other materials to the places where you found them.

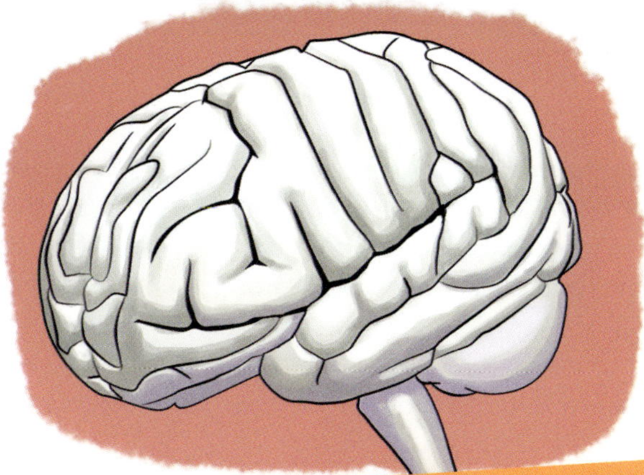

Sentient Science Skills

What Makes Some Memories So Memorable?

DIVE INTO THE FASCINATING SCIENCE OF MEMORY WITH A GROUP OF FRIENDS IN THIS QUICK ACTIVITY.

Have you ever tried to remember an entire list of groceries—without looking at the written list while you're shopping? Even a master of memory skills may fail to recall an item or two. But why does this happen? In this activity, you will learn a little more about memory—how it works and what factors make some details more memorable than others. You will a re-create a psychology experiment that helped scientists identify two effects that can distinctly shape what we remember.

PROJECT TIME

15 minutes

KEY CONCEPTS

Memory
Neuroscience
Psychology
Graphing

BACKGROUND

Can you recall all of the words in the paragraph on page 31? No peeking! Some memories are stored for only short periods of time before they disappear. You need to remember the beginning of a sentence to understand the end, for example, but might not need to retain each sentence word-for-word to understand the whole paragraph. This briefly held form of recollection is called short-term memory. Other memories, however, last much longer. Scientists believe that a seahorse-shaped structure in the brain called the hippocampus helps us transform our memories from short- to long-term.

Another form of recollection is working memory. This is what helps you remember and juggle information that is involved in tasks in which you are engaged. This is the form of memory that lets you perform mental math, conjugate verbs, or put together a list. Scientists have found that an area of your brain right behind your temples, called the prefrontal cortex, helps you carry out working memory tasks.

MATERIALS

- Pencils
- Paper, preferably lined or graph paper
- Watch or clock (optional)
- Gather a group of friends and family to be your subjects—at least five people (but the bigger the group, the better your results)

PREPARATION

- Give each of your subjects a piece of paper and pencil.
- Ask your subjects to listen carefully as you list 20 words. They will need to remember as many as they can but not write anything down until you tell them to.

PROCEDURE

- In a steady, even pace—about one word every second if you are using a clock or watch—read the following list of words in order: home, dog, rock, hand, table, card, bag, monkey, phone, cookie, mouse, paper, nail, hat, pillow, water, juice, watch, circle, and glass.

- *Why do you think your pacing matters? What do you think might happen if you repeated this activity and read more slowly? What if you read more quickly?*
- When you have finished reading the list, pause for one more second and then ask your subjects to write down as many words as they can remember.
- Collect their lists. On a fresh piece of paper, prepare a line graph. The vertical y-axis of the graph will represent the number of subjects in your study. Starting at the bottom of this axis, write the number 1, then evenly space consecutive numbers until you have reached the total number of subjects in your experiment. If five friends participated, you will have the numbers 1 through 5 going up this axis.
- The horizontal x-axis of your graph will include the words on your original list. Write these out, evenly spaced, in the same order that you read them. Your line graph will now allow you to pinpoint how many subjects remembered each of the words on the list.
- Count up how many subjects correctly remembered each word. Above each word on your graph's x-axis, mark a dot that corresponds with the number of subjects along the y-axis. *Are there certain words that everyone remembered? Are there words that all of your subjects forgot? Do some people have much sharper memories than others?*
- When you have placed a dot for each word, connect these points. *Do you notice any patterns? Does your line graph have a smooth shape or is it spiky? Why do you think certain words were more memorable than others?*

SCIENCE FAIR IDEA

In the original list, all of the words are simple and concrete, making them easier to imagine and relate to. Repeat this experiment with a new list of short, common words but include some that are idea-based, or abstract. You could also read a list of real and made-up words. *Are certain words harder to remember than others?*

EXTRA

Distract your subjects by adding a working memory challenge at the end of the experiment. Repeat the above activity with a new list, but before your subjects write anything down, ask them to recite the last 10 letters of the alphabet backward. Then let your subjects write down as many from the list as they can remember. Graph and compare your results. *Does disrupting memory change the pattern of remembered words?*

SCIENCE FAIR IDEA

You can also try to enhance memory. First, strategize with your subjects on good memory techniques. For example, many people find that visualizing something makes it easier to remember. Try a few examples as a group, then repeat the experiment with a new set of words. *Did your subjects remember more words with a little coaching?*

OBSERVATIONS AND RESULTS

Did most people remember the first and last words on the list but forget those in the middle?

Most people who try this experiment find their results create a U-shaped line graph. That means people do a good job remembering the beginning and end of the list but struggle to recall words in the middle. Two different memory effects create this pattern. The primacy effect suggests that we are good at remembering the very first information we encounter. Scientists are still unsure exactly how this effect works, but one theory is that individuals trying to remember words will repeat the growing list each time a new term is added. As a result, they repeat the first few items more than any other and this repetition shifts information from the short-term to more secure long-term memory. The second effect is the recency effect. This suggests it's easier to remember what we learned last because it is still fresh in our minds. If these theories are correct, your working memory, which is trying to reassemble the list of words, struggles to recall terms in the middle because they haven't yet been stored in long-term memory and have been pushed out of short-term memory by more recent additions.

If you've ever written a book report, you may have encountered both of these effects—it's easy to remember how a book ends and you can probably recall how it began, but the middle gets muddled. Luckily, once you have recognized these effects, you can also find ways to overcome them. In the case of that book report, taking extra notes is a simple solution that helps your memory keep details straight.

Recency and primacy are not the only effects that can influence your memory. You may have noticed that some words from the middle of the list were still very memorable. This could be because the words had special significance. For example, if a phone rang as you read "phone" or your subjects were very hungry when you said "cookie," these words could have gained added meaning.

CLEANUP

Return all materials to the places where you found them.

How Preference Affects Quick Choices

USE CANDY TO DETERMINE IF YOUR PERSONAL COLOR PREFERENCES DETERMINE THE CHOICES YOU MAKE.

Have you ever wondered if your preferences affect the choices you make? You make choices all the time, whether you are aware of them or not. Many different factors probably work together to affect the option you end up choosing. Which do you think are the most important?

Everyone has personal preferences, and these may affect the choices we make. What about your favorite color? You may prefer one color over another for no good reason, other than the fact that you like it—or you may have a personal reason for choosing a certain color. Do you think a color preference will affect your choices?

PROJECT TIME

20 minutes

KEY CONCEPTS

Psychology
Choices
Colors
Behavior

BACKGROUND

Might color preferences have biological origins? When we see a color, it is interpreted in our brain by the visual cortex, where different groups of neurons are stimulated. These different stimulations of neurons within the visual cortex might lead to color preferences, which might, in turn, affect our behavior.

While processing visual information, such as the perception of different colors, our brains also coordinate our muscle movements. This occurs in the brain's motor cortex. The visual cortex and motor cortex in our brains must function well together to allow for proficient hand-eye coordination. For example, when you catch a fast-moving ball, your eyes must tell your brain where the ball is, and then the brain tells your arm and hand muscles to make movements to catch it. Because the motor cortex must work so closely with the visual cortex, visual preferences, such as a favorite color, may affect movements and behaviors.

MATERIALS

- Bag of M&M'S (at least 12.6-ounce size)
- An extra volunteer or two participants (the more the better)
- Paper
- Pencil or pen
- Bowl

PREPARATION

- Pour out the bag of M&M'S and count out 50 of each color.
- Combine 50 M&M'S of each color in a bowl. Mix them together.

PROCEDURE

- Place one arm behind your back and pick out candies as quickly as possible using only a two-finger pinch. As the candies are pulled out, place the selected ones on the table next to the bowl. Stop when you have pulled out 20 candies.

- Write down the name of your favorite color. Then write down the total number of each M&M'S color that you pulled out. Return the pulled M&M'S back to the bowl. *Was the M&M'S color you chose most the same as your favorite color?*

- To make this activity even more scientific, repeat the procedure with other volunteers—and do not tell them what you are testing until after they have selected their candies from the bowl.

- Write down everyone's color preference and the number of each color of candy that they selected. If their favorite color is one that is not in the M&M'S bowl, ask them instead which of the colors of the M&M'S they like best.

- For each favorite color group, calculate the percentage of M&M'S chosen of each color. For example, if three participants said blue was their favorite color, and together they pulled a total of 18 blue M&M'S, out of a total of 50, then you would calculate that 36 percent of M&M'S picked by this group were blue. *If people have no color preference, and there is approximately an even distribution of the six colors of M&M'S in the bowl, what percentage of the time would you expect each color to be selected?*

- *Did your participants tend to pick their favorite color? Within each favorite color group, was there a color that was the second-most frequently picked one? What was the most frequently picked color overall?*

EXTRA

Try this experiment to test how preference affects choice in younger children. Ask a small child to choose a stuffed animal from an assortment of stuffed animals. Then ask what their favorite real animal is. Repeat this with multiple small children. *How many times does their favorite animal match the stuffed animal they chose?*

EXTRA

You can also try experiments to test mind-body coordination with handedness. *Will people tend to use the same hand with which they write to pick something up? Will people tend to step out first with the same side that they write with?* To make these activities the most reliable, explain what you are testing only after volunteers have completed the procedure—otherwise they might change their behavior based on what they think they are expected do.

OBSERVATIONS AND RESULTS

Did participants tend to pick their favorite color? Was one M&M'S color picked much more frequently than the others.

Preferences affect the quick choices we make all the time. What our favorite color is might be no exception. You might have seen participants tend to pick M&M'S that matched their favorite color. However, if a participant had chosen a certain color as their favorite not because the participant thought it was the most attractive color, but for other reasons such as a personal experience, the participant might not tend to pick M&M'S that matched their stated favorite. In addition to favorite color, other factors may also affect color selection, such as colors preferred in food, temperature, or lighting. What other considerations do you think affect your choices?

CLEANUP

Return all materials to the places where you found them. You can eat your M&M'S if you used clean hands to touch them!

Speaking Science

Why People Don't Hear What You Say

INVESTIGATE HEARING AND LISTENING BY PLAYING A FUN GAME WITH FRIENDS IN THIS ACTIVITY.

Have you ever told a friend or family member something only to later find that they completely misunderstood you—or never heard you at all? People often tell each other about important information that is not properly received, even when the conversation occurs in a quiet setting at close range. Why does this happen? In this activity, you will learn why communication can be so difficult by probing the psychology of listening. You will also experience how much a simple spoken message can be distorted.

PROJECT TIME

15 minutes

KEY CONCEPTS

Psychology
Attention
Working memory
Communication

BACKGROUND

The act of listening seems simple enough: the ears register the sounds produced and the brain interprets them, assuming the sounds reach the ears, and the listener knows the meaning of the words. In the real world, however, the situation is usually far more complicated. First, to mentally process the message, the person to whom you are speaking has to be paying attention. Not only may external distractions—a baby crying or a TV on in the background—divert their minds away from the words, but their own thoughts might also similarly lead them astray. Lost in thought, they are just not hearing you.

Processing language takes a fair amount of thought. We use a short-term mental sketch pad, so-called working memory, to hold each word and its meaning in mind long enough to combine it with others. If the meaning of any of the words is unclear, the task becomes harder.

In addition, people often don't express themselves clearly in the first place. They forget to include important background or context, which can dramatically shape the meaning of words. They might mumble some words or simply choose the wrong word, one that does not truly reflect what they mean. In short, even when the other person's brain is ready to listen, the information they need from a speaker is often not all there.

MATERIALS

- Pencil or pen
- Four or more friends or family members
- Stereo, TV, or other device that makes noise (optional)
- Paper

PREPARATION

- Ask your participants to sit in a circle.
- Explain that you will be whispering a statement to one person, who should then repeat it to the next person—and so on, until the message has made it around to everyone. Tell your participants to whisper the statement to their neighbor only once.

- Using a pen or pencil, write down a sentence that is at least 12 words long. For instance: "Sam dashed quickly over to Jimmy's house to remind his friend about the English assignment."
- Memorize the sentence; then fold over the paper, hiding your work. Don't show what you wrote.

PROCEDURE

- Join your friends in the circle.
- Whisper the sentence you memorized to the person next to you so that only that person can hear it.
- Ask your neighbor to whisper the statement to the next person in the circle.
- Observe the message make its way to every person in the group. **Notice:** *What else is going on around you? Is everyone paying attention equally well?*
- When the last person has heard the statement, ask them to repeat it out loud.
- Write that sentence on your piece of paper.
- Read your original sentence to the group.
- Compare the two sentences. *Are they different? Is the meaning of the distorted statement the same as the original or did it change in significant ways? If you really wanted to say this to someone, how much would the differences matter?*
- Ask the group if they could hear the statement clearly. *Did people generally think they got it right? What does their answer suggest about communication and miscommunication?*

EXTRA

Write a simpler, shorter sentence and repeat the activity. Compare the before and after sentences again. *Are they more similar this time? Why? Does your new result suggest anything to you about how to communicate more effectively?*

EXTRA

Create another more complex statement like your original one. This time, let people say the sentence to their neighbor more than once. Compare the before and after sentences again. *Are they more similar this time? Why? Does your new result suggest anything to you about how to communicate more effectively?*

EXTRA

Try the same activity again with a new simple sentence, but this time with a radio or TV on in the room. *Do people look more distracted? How close is the final sentence to the one you started out with? Do you think having the background noise made it more difficult to pay attention to and remember the sentence?*

EXTRA

Ask your group to talk about times in which they remember being misunderstood. *What happened? What were the consequences? What do they think went wrong in those cases?* Talk about the best practices for good spoken communication. *Can you come up with five rules?*

OBSERVATIONS AND RESULTS

Did the statement you whispered change a lot during its journey? Very often the differences from the original are so great that people laugh. In real life, of course, miscommunication can be more frustrating than funny. This frustration might come from the belief that communication is straightforward, even though, as we have learned in this activity it is not.

The world is full of distractions, both external and internal. No one can control all of them at any given moment. Moreover, everybody's brain is different—in how it works and in the information and experiences it has collected. Thus, what you think you are saying may mean something quite different to someone else—particularly if you start in the middle of a thought, choose a wrong word, or speak too quickly. Speakers make one or more of these mistakes quite often—and worse, rarely realize when they do.

Once you realize the obstacles to communication, you will be far more understanding when it fails—as well as able to communicate more effectively by averting common errors. Think about what is going on at the moment: Does the other person appear to be distracted? If you need to repeat yourself, don't be annoyed. Repetition, as you may have learned from the above activity, is a good strategy for making sure you are understood. Another tactic is simplicity. See if a bare-bones message will suffice, at least for now, but don't leave out background or critical details!

CLEANUP

Return all materials to the places where you found them.

Shape Science
The Bouba-Kiki Effect

DID YOU KNOW THAT MADE-UP WORDS CAN TEACH YOU A LOT ABOUT LANGUAGE? TRY THIS ACTIVITY TO LEARN HOW!

Do shapes have certain "sounds" to people, regardless of what language people speak? For example, does everyone associate certain physical characteristics, like sharpness and roundedness, with certain sounds? Understanding such similarities may not only help people better understand how languages develop, but it may also improve people's ability to communicate, especially when trying to cross language barriers. In this activity you'll investigate the "Bouba-Kiki effect" to find out how abstract shapes may be linked to sound.

PROJECT TIME

15 minutes

KEY CONCEPTS

Language
Speech
Shapes
Sound
Psychology
Human behavior

BACKGROUND

One of the most amazing things humans do is use language to communicate. People have developed languages over many tens or hundreds of thousands of years, resulting in many very different tongues being spoken around the world today. How did our ability to use language evolve? Where and when was it first used?

One idea is that the first languages were representative sounds that people linked to concepts. Eventually these meaningful sounds grew more diverse and structurally intricate as they evolved into more complex speech. This idea brings with it a major question: Were the first sounds made by humans arbitrary and random, or were they consistently applied to certain symbols and concepts? A psychological phenomenon called the Bouba-Kiki effect has been used to help answer this question. In this effect, people are shown a pointy picture or a curvy, bubbly shape and asked to identify it as "Bouba" or "Kiki," even though those are both nonsense words. They are asked to do the same with a pointy, sharper shape. A surprising number of people identify the round shape as "Bouba" and the pointy shape as "Kiki" although they have not been told what the words mean. Will you observe the Bouba-Kiki effect or will you have the results of a 50-50 random-chance event?

MATERIALS

- A pen, pencil, or marker. You can use a colorful assortment if you want!
- 20 index cards
- A piece of paper
- At least five volunteers

PREPARATION

- On 10 index cards, draw a pointy, star-like shape.
- On the other 10 index cards, draw round, bubbly shapes.
- Randomly mix the cards by shuffling all of them together a few times.

PROCEDURE

- Tell your first volunteer that you will show them a series of cards, and that you want to know if the picture is Bouba or Kiki, but do not tell them what Bouba or Kiki mean.
- If they ask you for a definition, just explain that they are supposed to guess and do their best to decide.
- Show your volunteer the cards, one at a time.
- On a piece of paper keep track of the number of "correct" and "incorrect" answers. "Correct" answers are when the volunteer identifies a pointy, starlike shape with Kiki or a round, bubbly shape with Bouba. (Note that the answers your volunteers give you are not actually correct or incorrect, because "Bouba" and "Kiki" are made-up words.)
- *How many correct answers did the volunteer give? Did the volunteer give any incorrect ones, matching Kiki with the round shape or Bouba with the pointy shape?*
- Repeat these steps with at least five more volunteers. Use the same deck of index cards each time but shuffle it so that the cards are randomly mixed.
- *When you finished, what was the total number of correct responses and the total number of incorrect responses from all volunteers? Did the volunteers mostly give correct responses?*

SCIENCE FAIR IDEA

Try to find volunteers from a series of different age groups and do this activity again. *Does the Bouba-Kiki effect differ by age group? What is the youngest group in which you can observe the effect?*

EXTRA

Give a new volunteer a piece of paper and ask the volunteer to draw something that is Bouba or Kiki. Try this with several new volunteers, collect all of the drawings, and score them for being pointy or rounded in shape. *How well does this work? Does the volunteer always make a Bouba drawing rounded and a Kiki one pointy?*

EXTRA

If you know people who speak different languag try this activity on them. *Does it still work?*

EXTRA

What is it about the words "Bouba" and "Kiki" that make this work? Try this activity again but this time swap the vowel sounds by using the words "Bee-Bee" and "Kooka" instead of "Bouba" and "Kiki" when volunteers look at the drawings. *Are the results the same, or do your results switch around?* If they're the same, then perhaps the consonant sounds of the "B" and the "K" are most important, but if they're switched, then the vowel sounds of the "ee" and "oo" may be more important.

OBSERVATIONS AND RESULTS

Did all of your volunteers identify the round shape as Bouba and the pointy shape as Kiki, even though they were not told what the words mean? Did you record no incorrect answers?

You most likely found that every volunteer you tested gave only correct answers, or almost entirely correct answers. In other words, the volunteers always, or nearly always, identified each round shape as Bouba and each pointy shape as Kiki. People who have different native languages make this association, as do even very young children most of the time. What does this mean?

Is this evidence of a human predisposition to associating certain sounds with abstract concepts? This could indicate that humans did indeed apply the first sounds made in burgeoning languages to certain concepts or symbols, and that the associations with sounds in different languages are not actually random. One theory is that Bouba is associated with the round shape because a more rounded shape is made by our mouths when forming this sound whereas our mouths make a more angular shape when we say "Kiki." Additionally, "K" is a harder-sounding letter than "B." Those who use the modern Latin alphabet (including English speakers) may also be visually swayed: the written letters "K" and "B" are sharp and rounded, respectively.

CLEANUP

Return all materials to the places where you found them.

Savory Science
Jelly Bean Taste Test

LEARN HOW YOUR SENSE OF SMELL INFLUENCES THE WAY YOU TASTE FOOD IN THIS DELICIOUS EXPERIMENT.

Thanksgiving brings a feast of flavors. But when you imagine the mouthwatering meal—the tang of ruby-red cranberry sauce or sweet, cinnamon-scented pumpkin pie—you might notice that you are combining sensory cues. Clearly the senses work together in your recollection, but how much is taste influenced by other sensory information as you eat? In this activity, you'll find out by looking at two senses in particular.

PROJECT TIME

15 minutes

KEY CONCEPTS

Taste
Olfaction
Gustation
The five senses

BACKGROUND

Every time you take a bite of food, receptors in your mouth called taste buds pick up the taste of the food you are eating. These receptors are sensitive to five basic tastes: umami (a savory flavor), salty, sweet, bitter, and sour. But right above your mouth is your nose, which also plays a part in how you experience food. The nose is equipped with millions of receptors for odor molecules. You can smell food by sniffing through your nostrils or if air is circulating inside your nose as you chew. The latter occurs because the back of your throat connects your nose and mouth. The only catch is that air needs to be flowing in or out of your nose for the odor molecules to get into the nose—either through the front or the back. This explains why pinching your nose prevents you from smelling food.

Once they arrive in your nose, odor molecules travel to your nose's olfactory epithelium, the area of the nasal cavity where odor detection occurs. While you are eating, your brain receives signals from both your mouth and nose, allowing you to recognize whatever tasty treat you happen to be chewing. In this activity, you'll separate the sensations of taste and smell to learn how much each contributes to your recognition of a familiar food.

MATERIALS

- Jelly beans (at least three different fruit flavors work best)
- Pencil and paper
- Plastic sandwich bags
- A partner. You can also work with a group of friends and compare results.
- Blindfold (optional)
- Glasses of water (optional)

PREPARATION

- Separate your jelly beans by flavor.
- Select at least three plastic bags—one for each flavor you want to use in the experiment. Place a few appropriately flavored jelly beans in each bag. For example, one bag could be for mango jelly beans, another for strawberry, and a third for banana-flavored ones. Push down on the bags to smush the candies slightly.

PROCEDURE

- Ask your partner to close their eyes (or use a blindfold).
- Give your partner a jelly bean. Ask him or her to chew it and guess its flavor. Record the response, along with the correct answer. Repeat with at least two other flavors. You can offer your subject a glass of water between samples to clean their palate. *How good is your partner at guessing the jelly bean's flavor?*
- Tell your partner to pinch their nose shut, then hand your partner a jelly bean. Ask them to eat the candy and tell you what flavor they taste. Record the response along with the correct answer.
- Repeat the previous step with one or two other jelly bean flavors—you can offer your subject a sip of water in between each to cleanse their palate. Record each response, along with the correct answer. *Does being unable to smell change your subject's responses?*
- With eyes still shut or blindfolded, ask your partner to breathe deeply while you open one of the plastic bags that hold crushed jelly beans. Ask your subject to guess which flavor they are smelling—record the response and correct answer. Repeat with the other two bags. *Is your subject better at guessing based on taste alone or scent alone?*
- Switch roles with your partner or repeat the above with another subject. *Is it easy to recognize the jelly bean flavor by taste? By scent? How do your results compare with your partner's?*

EXTRA

You can also try this on a larger group of people and see whether certain groups are better than others under different conditions. *For example, are older people better or worse at guessing the candy's flavor by scent alone?*

EXTRA

For a variation on this experiment, peel, and chop two potatoes and two apples. (Have an adult supervise when you use knives or peelers.) You'll notice the peeled apple and potato slices look very similar. Hand either a slice of apple or potato to a partner—don't let your partner know which is which. Ask your partner to take a bite while keeping their nose pinched closed. *Can your partner tell the difference between the apple and the potato? Can you?* Try again with nose un-pinched.

SCIENCE FAIR IDEA

In this experiment, subjects can't see the jelly bean's color, but if you want to check whether vision influences how something tastes, set up a soda-tasting experiment. Get three kinds of fruit soda—such as cherry, grape, or orange—and one flavor-free carbonated water. Add a few drops of food coloring to the carbonated water. (Try to use a color that differs from the sodas.) Pour your beverages into glasses and ask subjects to taste each one and name the beverage's flavor. *Do subjects mistake the colored water for another fruit soda? Does the color of the water trick people into expecting a soda flavor to match the color?*

OBSERVATIONS AND RESULTS

Did plugging your nose make it difficult to distinguish a jelly bean's flavor? Could you and your partner recognize a flavor by sniffing the crushed candies?

When you cannot smell the jelly bean you are eating, you can only taste the candy's sweetness—and that's not enough information to tell which flavor you are chewing. This demonstrates how much we rely on our sense of smell when we "taste" food—much of the experience comes from scents rather than taste itself. This is also why everything tastes bland when you have a cold: Your stuffy nose keeps you from enjoying the full olfactory experience. In addition to scent and taste, other factors including a food's temperature and texture affect how you experience and interpret each bite.

CLEANUP

Return your materials to where you found them. You can eat the leftover jelly beans if you choose!

THE SCIENTIFIC METHOD

The scientific method helps scientists—and students—gather facts to prove whether an idea is true. Using this method, scientists come up with ideas and then test those ideas by observing facts and drawing conclusions. You can use the scientific method to develop and test your own ideas!

Question: What do you want to learn? What problem needs to be solved? Be as specific as possible.
Research: Learn more about your topic and refine your question.
Hypothesis: Form an educated guess about what you think will answer your question. This allows you to make a prediction you can test.
Experiment: Create a test to learn if your hypothesis is correct. Limit the number of variables, or elements of the experiment that could change.
Analysis: Record your observations about the progress and results of your experiment. Then analyze your data to understand what it means.
Conclusion: Review all your data. Did the results of the experiment match the prediction? If so, your hypothesis was correct. If not, your hypothesis may need to be changed.

GLOSSARY

audition: The sense of hearing.
cognitive: Having to do with to thinking.
complex: Made of many parts.
conjugate: To give different forms of something.
consecutive: Following one after the other, in order.
disorient: To lose sense of direction or be confused.
distinguish: To tell the difference between things.
distort: To pull out of shape.
divert: To change course.
engaged: Busy or occupied.
enhance: To make better.
gustation: The sensation of tasting.
magnitude: The size of something compared with something else of the same kind.
pace: Consistent and continuous speed.
proficient: To be very good at something.
receptor: An organ or cell able to respond to light or heat and transmit a signal.
variables: Things that can change.
vigorous: Full of energy.

ADDITIONAL RESOURCES

Books

Calandrelli, Emily. *Stay Curious and Keep Exploring: 50 Amazing, Bubbly, and Creative Science Experiments to Do with the Whole Family.* San Francisco, CA: Chronicle Prism, 2022.

Dickinson, Michelle. *Kitchen Cabinet Science Projects: Fifty Amazing Science Experiments to Make with Everyday Ingredients.* New York, NY: Penguin Workshop, 2021.

Heinecke, Liz. *Sheet Pan Science: 25 Fun, Simple Science Experiments for the Kitchen Table.* Beverly, MA: Quarry Books, 2022.

Scalzo Yi, Andrea. *Super Simple Science Experiments for Curious Kids: 100 Awesome Activities Using Supplies You Already Own.* Salem, MA: Page Street Publishing Co, 2022.

Websites

Discovery Education
sciencefaircentral.com

Exploratorium
www.exploratorium.edu/explore/activities

Science Buddies
www.sciencebuddies.org/science-fair-projects/project-ideas/list

Science Fun for Everyone!
www.sciencefun.org/?s=science+fair

Videos

Learn How Experiments are Conducted on the ISS | Smithsonian Science Starters
https://wtvi.pbslearningmedia.org/resource/learn-how-experiments-are-conducted-on-iss-video/smithsonian-science-starter/

How to Make a Bottle Thermometer | Full-Time Kid
https://wtvi.pbslearningmedia.org/resource/how-to-make-a-bottle-fulltimekid/how-to-make-a-bottle-fulltimekid/

Grilled Cheese Stackers | Mission: Materials Science
https://wtvi.pbslearningmedia.org/resource/grilled-cheese-stackers-video-gallery/grilled-cheese-stackers-mission-materials-science-video-gallery/

Publisher's note to educators and parents: Our editors have carefully reviewed these websites to ensure that they are suitable for students. Many websites change frequently, however, and we cannot guarantee that a site's future contents will continue to meet our high standards of quality and educational value. Be advised that students should be closely supervised whenever they access the internet.

INDEX